An introduction to reliability engineering

Consulting Editor:
M. R. Ward
Vice-Principal, South East London Technical College

Other books in the Technical Education Series

Hutson: Colour Television Theory: PAL-system Principles
and Receiver Circuitry
Morris: Control Engineering
Morris: Industrial Electronics: for Technicians and
Technician Engineers
Smithson: Mathematics for Electrical and Telecommunications
Technicians, Volumes 1, 2, and 3.
Stott and Birchall: Electrical Engineering Principles: for
Electrical, Telecommunications and Installation Technicians
(revised first edition)
Ward: Electrical Engineering Science

An introduction to reliability engineering

Rhys Lewis, B.Sc. Tech., C.Eng., M.I.E.E.
Llandaff Technical College, Cardiff.

 McGRAW-HILL

London · New York · Sydney · Toronto ·
Mexico · Johannesburg · Panama · Düsseldorf

Published by
McGRAW-HILL Publishing Company Limited
MAIDENHEAD · BERKSHIRE · ENGLAND

---

09 094263 3

PRINTED IN GREAT BRITAIN

Preface

The subject of reliability and reliability engineering is
playing an ever increasing role in present day technology,
its importance taking on a new level with the demand for
an ability to predict operational behaviour of systems as
intricate as those which recently put man on the moon.
It is with this in mind that the City and Guilds of London
Institute have now incorporated an introduction to the
subject in their course 57, fourth year electrical technicians syllabuses.

Reliability engineering is now a fine art, is complex,

and demands considerable knowledge of statistical mathematics. This short textbook is written specifically for those who do not have that knowledge. Wherever possible physical explanations have been included alongside the mathematical treatment, in an attempt to help the reader to see *why* certain behaviour takes place, rather than to accept blindly the fact, shown mathematically, that it does.

The text is intended particularly for those students on the course indicated above and the contents are in accordance with the official syllabus. However, it should also prove useful to practising engineers to whom, until now, reliability has been a somewhat vague and imprecise term.

RHYS LEWIS

Contents

1. Failure and reliability

1.1 Introduction

Reliability is the characteristic of a component, or system made up of many components, expressed by the probability that it will perform its particular function within a specific environment for a given period of time. Since the subject of reliability is obviously concerned with statistical analysis and the prediction of behaviour based on tests, it might be considered, at first glance, to be a matter of guesswork or chance. However, reliability predictions have become a precise branch of industrial technology.

1

Reliability engineering plays an invaluable part in the reduction of costly failures and the correct planning of overhaul and maintenance schedules.

1.2 Types of failure

Failure is defined as the inability of a component or system to carry out its specified function. Failures may be categorized in a number of ways according to the degree of failure, the reason for failure, the timing of the failure and so on. The following are some relevant definitions:

Misuse failure due to overloading or otherwise overstressing a component or system beyond its capability.

Inherent-weakness failure due to inherent weakness of the component or system, and occurring while the item is being correctly used.

Sudden failure is used to describe failures which could not have been anticipated.

Gradual failure is used to describe failures which could have been anticipated.

Partial failure is one in which the component or system may still function, but not to the limits of performance originally designed.

Complete failure results in total loss of the required function.

Catastrophic failure is one which is both sudden and complete.

Degradation failure is one which is both gradual and partial.

Chance failures is a term used generally to describe those failures which occur suddenly and at random during the anticipated useful life of a component or system. They may be due to a variety of reasons including inherent weakness, misuse, etc.; they are not due to the component having completed its normally anticipated useful life, i.e., to the component wearing out.

2

Wearout failure is another general term to describe failure due to the wearing out of a component which has more or less completed its anticipated useful life.

Both chance and wearout failures may be partial or complete.

1.3 Failure rate

The number of failures occurring per unit time is known as the failure rate. As with all quantities describing change (speed, acceleration, etc.) an average value may be obtained by dividing the total number of failures which have occurred during a time interval by the length of the interval. The shorter the interval, the nearer the average value gets to the instantaneous failure rate. The *instantaneous* failure rate at any one time is the slope of the curve plotting failures against time at that particular time.

If, in determining the failure rate, the number of failures occurring during the time interval is expressed as a proportion of the number of survivors at the beginning of the time interval, then the failure rate obtained is called the proportional failure rate. It is denoted by the symbol λ. Unless otherwise stated the words 'failure rate' within this text imply the proportional failure rate.

Example

If out of 1 000 components 10 fail during a period of 5 000 hours then the

proportional failure rate
$$= \frac{10}{1\,000} \times \frac{1}{5\,000},$$

i.e.,
$$\lambda = 2 \times 10^{-6} \text{ failures of the total/hour}$$

and the

percentage failure rate
$$= \frac{10}{1\,000} \times \frac{1}{5\,000} \times 100$$
$$= 0 \cdot 000\,2 \text{ per cent/hour.}$$

3

Failure rate is most commonly expressed as a percentage per 1 000 hours. The above would then be

$$= 0.2 \text{ per cent}/1\,000 \text{ hours}.$$

1.4 The bathtub diagram

A typical graph plotting against time the percentage failure rate with respect to time is shown in Fig. 1.1. It is often referred to as the 'bathtub' diagram due to its shape.

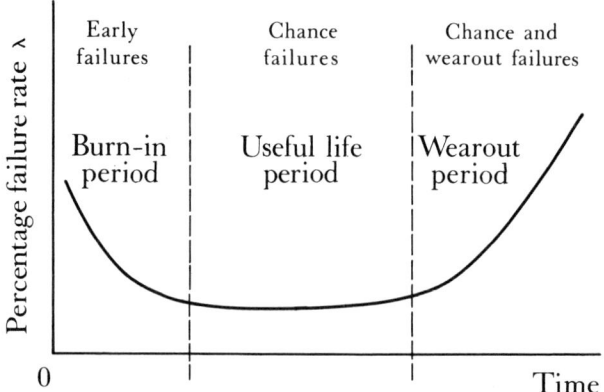

Figure 1.1 The bathtub diagram

During the burn-in period a high failure rate exists, due to the presence of substandard components in the sample tested. After the weak components have died out the failure rate stabilizes at an approximately constant value; this period is called the useful life period.

Eventually wearout failures begin to occur and the failure rate rises again. During this wearout period chance failures may, of course, still be occurring.

In order to make reasonably accurate predictions of reliability, failures due to chance and wearout must be studied and an analysis made of the times in relation to the type of failure.

4

1.5 Constant failure rate case

When the failure rate is constant, reliability prediction is made much easier mathematically, since it is possible to use exponential curves to assist analysis.

As was shown above, the failure rate may be assumed to be constant when failures are due to chance alone; this can be achieved by correct overhaul schedules, which eliminate wearout failure. It is also possible, however, to achieve a constant failure rate, and thus simplify the mathematics involved, by a process of immediate replacement on wearout. This latter case is not as obvious but has been conclusively demonstrated. It should be noted that replacement on failure is a procedure which cannot always be adopted, since certain systems or components cannot be allowed to fail even temporarily.

1.6 Reliability equations and curves when failure rate is constant

The probability of no failures occurring in a given time can be expressed by the following equation, provided that the failure rate is constant:

$$R = e^{-\lambda t}, \tag{1.1}$$

where R is the probability of no failures in time t, i.e., the reliability, e is the exponent 2·7183, λ is the constant failure rate. (This is in fact the no-event term of the Poisson probability function.) The *unreliability*, Q, is defined as the probability of total failure. It follows logically that

$$R + Q = 1, \tag{1.2}$$

and that

$$Q = 1 - e^{-\lambda t}, \tag{1.3}$$

where Q is the probability of total failure in time t. A

5

graph of R and Q against time yields the familiar exponential curves as shown in Fig. 1.2.

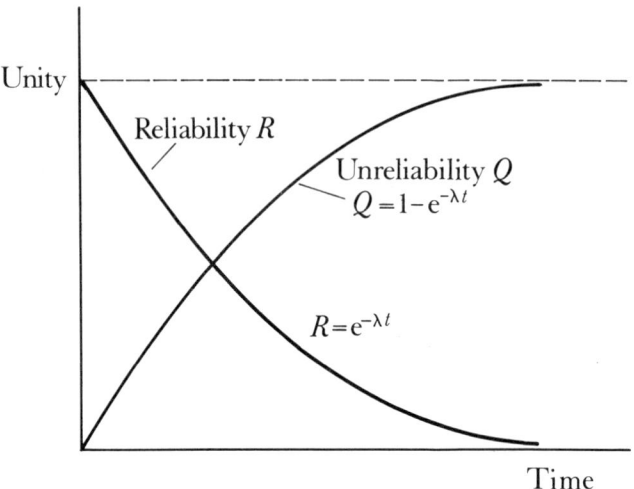

Figure 1.2 Reliability and unreliability curves

Notice that

at time $t = 0$, $R = 1$ and $Q = 0$,

at time $t = \frac{1}{\lambda}$, $R = 0{\cdot}37$ and $Q = 0{\cdot}63$

(from tables of values of e raised to various powers). A graph of survivors, i.e., the number of components still alive at time t against time, will yield the same shape as the reliability curve above. A graph of failures against time yields the same shape as the unreliability curve (see Fig. 1.3). The equation of the graph of survivors versus time (survival curve) is

$$N_s = N_0 \, e^{-\lambda t}, \tag{1.4}$$

where N_s is the number of survivors at time t, and N_0 is the original number in the test sample. Similarly the

6

equation of the failures, versus time graph (failure curve) is

$$N_f = N_0 (1 - e^{-\lambda t}), \qquad (1.5)$$

where N_f is the number of failures at time t, and N_0 is as before.

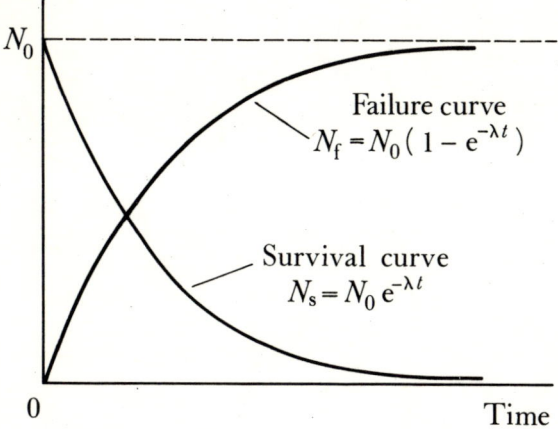

Figure 1.3 Failure and survival curves

Using these curves and the equations above, the reliability or unreliability, number of survivors or failures, may be calculated at any time, provided that the failure rate λ is known (*and* constant, since the exponential analysis only applies when this is so).

2. Failure intervals

2.1 Mean time between failure, MTBF

The mean time between failure, abbreviated MTBF, of a component or system is an extremely important characteristic in reliability predictions. It is defined as the mean or average time which elapses between failures, and usually refers to a situation in which the failure rate λ is constant, i.e. due to chance failures or the adoption of the replacement-on-failure technique described earlier. The symbol for MTBF in these cases is m.

To appreciate the theoretical derivation of m, we must

look more closely at the survival curve and the reliability curve and examine the significance of the area contained between the curves and the axes. Firstly, consider a survival curve which is not exponential, such as the one shown in Fig. 2.1.

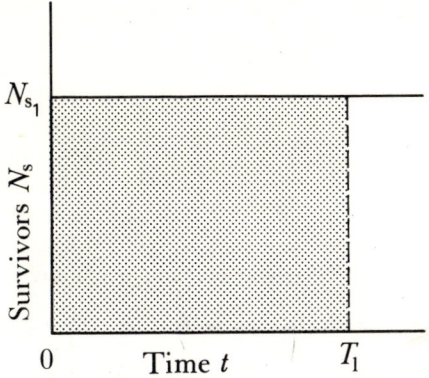

Figure 2.1 A nonexponential survival curve

This is in fact a very simple function indeed and shows a constant number of survivors, i.e., a zero failure rate. At time T_1 shown the survivors number N_{s_1}. Multiplying N_{s_1} by T_1 gives us the *survivor-hours*, i.e., the total number of hours of survival by all components. This value $N_{s_1} T_1$ is clearly the area bounded by the curve, the axes, and the line $t = T_1$. The area beneath a survival curve, whatever its shape, is in fact always equal to the total survival hours of all the components. To return to the exponential survival curve, which we have seen is obtained for a constant (nonzero) failure rate, the same principle applies. In this case, after a very long time (depending on the failure rate) all components N_0 have failed and the survival curve reaches the time axis. The area under the curve represents the total survival hours of all N_0 components. If we now divide this area by the total failures, i.e. N_0, this gives the average time between failures, the

9

MTBF, m. The area under a nonlinear curve such as this is, of course, obtained by integration and in this case

$$\text{MTBF} = \frac{1}{N_0}\int_0^\infty N_s\, dt$$

$$= \frac{1}{N_0}\int_0^\infty N_o\, e^{-\lambda t}\, dt$$

$$= \int_0^\infty e^{-\lambda t}\, dt$$

$$= \left[-\frac{e^{-\lambda t}}{\lambda} \right]_0^\infty$$

$$= 0 - (-1/\lambda) = 1/\lambda, \qquad (2.1)$$

and we find that the MTBF is, in fact, the *reciprocal* of the failure rate.

Since the survival curve is the same curve as the reliability curve, except that the vertical axis of the reliability curve has been multiplied by N_0 to give the survivors axis of the survival curve, then it follows that the MTBF is in fact equal to the area under the reliability curve divided by unity, i.e., equal to the area itself.

Since

$$m = 1/\lambda$$

the equations for reliability R and unreliability Q at any time can be rewritten as

$$R = e^{-t/m}, \qquad (2.2)$$

and

$$Q = 1 - R = 1 - e^{-t/m} \qquad (2.3)$$

and the equations for the number of survivors N_s or failures N_f at any time can be written

$$N_s = N_0\, e^{-t/m}, \qquad (2.4)$$

$$N_f = N_0\, (1 - e^{-t/m}), \qquad (2.5)$$

where N_0 is the number of components alive at the start of the test.

Notice that since $m = 1/\lambda$ and since, as was shown earlier, when $t = 1/\lambda$ the reliability has fallen to $0 \cdot 37$ and the survivors to $0 \cdot 37\ N_0$, the MTBF for the exponential case is the time at which

$$R = 0 \cdot 37 \qquad Q = 0 \cdot 63$$

$$N_s = 0 \cdot 37 N_0 \qquad N_f = 0 \cdot 63 N_0$$

2.2 Measurement of MTBF (chance failures)

The two main types of test for measuring m for chance failures are the non-replacement and the replacement. The latter is seldom used since it involves constant observation to ascertain exact moments of failure. Non-replacement methods require observation only at the beginning and end of the test time. To exclude wearout failures the test is truncated (cut off) before the wearout probability is too high. It has been demonstrated by Epstein that the best estimate of m for a truncated test is given by

$$m = \frac{\text{test hours for failures} + \text{test hours for survivors}}{\text{number of failures}},$$

that is,

$$m = \frac{\text{total component test hours (survival hours)}}{\text{total number of failures}}. \tag{2.6}$$

It will be appreciated that the figure obtained is only an estimate; the confidence which one can have in such an estimate may be determined, using standard statistical methods. These are briefly considered under 'confidence limits' in chapter 3.

For a constant failure rate the MTBF may also be determined by finding the reciprocal of the failure rate, as shown above.

11

Worked examples on the determination of failure rate, MTBF, reliability, and survivors, etc., using the formulae given so far, are provided in the following pages.

Examples

1. The telemetry transmission system of an earth satellite has an MTBF of 10 000 hours. Estimate the probability of no failures during 100 90-minute orbits.

$$m = 10\ 000\ \text{hours}$$
$$\text{time} = 150\ \text{hours,}$$

therefore probability of no failure, i.e., the reliability, is given by

$$e^{-150/10\ 000} = e^{-0.015}$$
$$= 0.8607.$$

There is therefore an 86 per cent probability of 100 failure-free orbits.

2. A certain electronic control system has a constant failure rate established at 0·2 per cent/1 000 hours. Determine the probability of 500 hours of failure-free operation.

$$\text{reliability for 500 hours} = e^{-(2/10^6)500}$$
$$= e^{-0.001}$$
$$= 99.88\ \text{per cent.}$$

Notice that the failure rate λ, since it equals 0·2 per cent/1 000 hours, is equal to

or
$$0.2\ \text{per 100 000 hours,}$$
$$2\ \text{per 1 000 000 hours,}$$

i.e., $2/10^6$

and it is this figure which is inserted into the reliability equation. More worked examples are included at the end of the text.

2.3 Wearout failures: mean wearout life

As has been shown, chance failures are distributed exponentially, approximately 63 per cent occurring before a time equal to the MTBF and approximately 37 per cent occurring afterwards. Failures due to wearout, i.e., to the component having completed its anticipated useful life, are not distributed in this manner. A graph of wearout failures against time has the shape shown in Fig. 2.2.

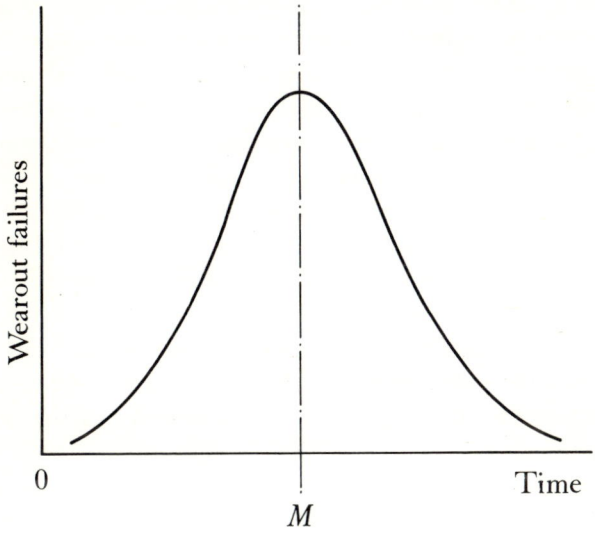

Figure 2.2 Wearout failure curve

This type of distribution is well known in statistical analysis; it is called the *Gaussian* or *normal* distribution and has certain defined characteristics which clearly distinguish it from other distributions.

The failures due to wearout are clustered about an average or *mean* value of time. Since each point on the curve corresponding to, say, x failures occurring at time t means that x of the components failed after a useful life t, the mean value M corresponds to the *mean wearout life* of the components. The individual lives of the components are scattered *normally* about the mean. Mean wearout life must not be confused with mean time between chance failures, discussed above. The M T B F (chance) tells us the average time anticipated between chance failures *during* the useful life of a component or system; the mean wearout life, on the other hand, tells the average value of the anticipated useful life assuming failure due to chance does not occur. Suppose, for example, a component has a value of M equal to 10 000 hours; its mean time between chance failures (computed from a test which is truncated long before 10 000 hours) may be as high as 100 000 hours. These two figures indicate that provided the component is used within the useful life period, i.e., up to 10 000 hours, the probability of chance failure using the formulae described above and a failure rate λ of $1/m$, i.e., $1/100\ 000$ or $0{\cdot}000\ 01$, is quite low. After the anticipated useful life is over the probability of failure rises very rapidly due to a very much increased failure rate, which is, itself, due to wearout beginning to take place. Wearout and chance failures can be distinguished from one another by careful examination of the physical characteristics of the dead component.

An important parameter concerned with any normal distribution is the *standard deviation* σ. This is computed by finding the square root of the mean of the square of the deviations of the measured characteristic from the average value.

That is,

$$\sigma = \sqrt{\dfrac{\text{sum of squares of deviations from average}}{\text{total number of observations}}} \quad (2.7)$$

or, for wearout tests, if the life of components 1, 2, 3, ..., n

is indicated by $t_1, t_2, t_3, \ldots, t_n$, and the total number used is n, then

$$\sigma = \sqrt{\frac{(t_1 - M)^2 + (t_2 - M)^2 + \cdots + (t_n - M)^2}{n}}, \quad (2.8)$$

where M is the mean (wearout) life.

For a normal distribution of failures it can be shown that approximately 68 per cent of the failures occur within a period $M \pm \sigma$, i.e., between time $M - \sigma$ and time $M + \sigma$, approximately 95 per cent occur within a period $M \pm 2\sigma$ and 99·7 per cent occur within a period $M \pm 3\sigma$. This is useful when establishing the confidence that one can have in estimates of wearout life (or indeed of any variable which has a normal distribution). See Fig. 2.3.

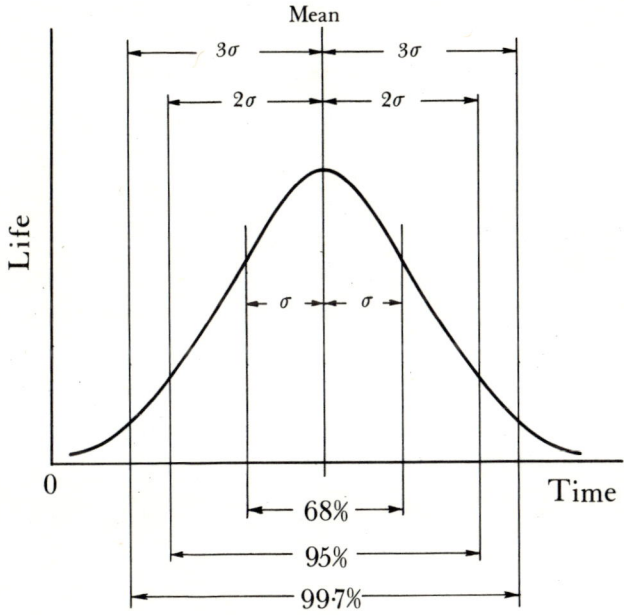

Figure 2.3 Percentage points on the normal distribution curve

2.4 Measurement of mean wearout life

For a wearout life test, a sample of components is put on test under the conditions they will experience in service, and the test is run until the components fail. Careful examination of both physical characteristics of the dead components and of their life length eliminates chance failures. (It is clear, for example, that if a group of components has lives centring around, say, 10 000 hours then a component surviving only 1 000 hours is probably not a wearout failure.)

The mean life is then computed as follows:

$$M = \frac{\text{sum of lives of components}}{\text{number of components}}. \qquad (2.9)$$

The standard deviation may be determined using the equation already given.

3. Confidence limits

3.1 Confidence limits

As can be seen, the two important parameters in reliability prediction—the MTBF m (when the failure rate is
constant) and the mean wearout life M—are only estimates and may lie either side of the true values of m and
M respectively. Clearly, the greater the number of tests,
the closer will be the estimates to the true values. The
question arises, when estimates are made, as to how close
they are to the truth; i.e., what confidence can be placed
in the estimate values?

Statistical theory again plays a large part in the assessment of the confidence level, and 'confidence' is usually expressed as the probability that the true value lies within a certain range, called the *confidence interval*, between two levels or limits called the *lower* and *upper confidence limits*. These limits are determined by the application of standard statistics theory to the estimated value.

3.2 Confidence limits for normal distributions

We have already seen that for a normally distributed parameter (such as mean wearout life) that about 68 per cent of the values occur within the range (mean $\pm \, \sigma$)
    about 95 per cent of the values occur
                within the range (mean $\pm \, 2\sigma$)
    and 99·7 per cent of the values occur
                within the range (mean $\pm \, 3\sigma$)
where σ is the standard deviation of the wearout life distribution. When a number of tests is run to determine the mean or average wearout life, several possible values of the mean life will be obtained. These are, in fact, *estimates* of the *true mean*. It can be shown that since the original characteristic, i.e., wearout life, has a normal distribution about the mean wearout life, then the estimates of the true mean will also have a normal distribution, this time about the *true* mean wearout life. The standard deviation σ of the life distribution will not be the same as the standard deviation for the means distribution. It can be shown that the means distribution will in fact have a standard deviation given by:

$$\frac{\text{standard deviation of wearout life distribution}}{\sqrt{\text{number of components tested to failure}}},$$

where these figures are obtained from a single test. Thus for a wearout life distribution standard deviation σ and for n components tested, then

standard deviation of means distribution $= \sigma/\sqrt{n}$, (3.1)

and we can say that

about 68 per cent of the values of the mean
will lie in the range $\hat{M} \pm \sigma/\sqrt{n}$,

about 95 per cent of the values of the mean
will lie in the range $\hat{M} \pm 2\sigma/\sqrt{n}$,

and 99·7 per cent of the values of the mean
will lie in the range $\hat{M} \pm 3\sigma/\sqrt{n}$,

where $\hat{M}$ is the *true mean wearout life*. Examine Fig. 3.1.

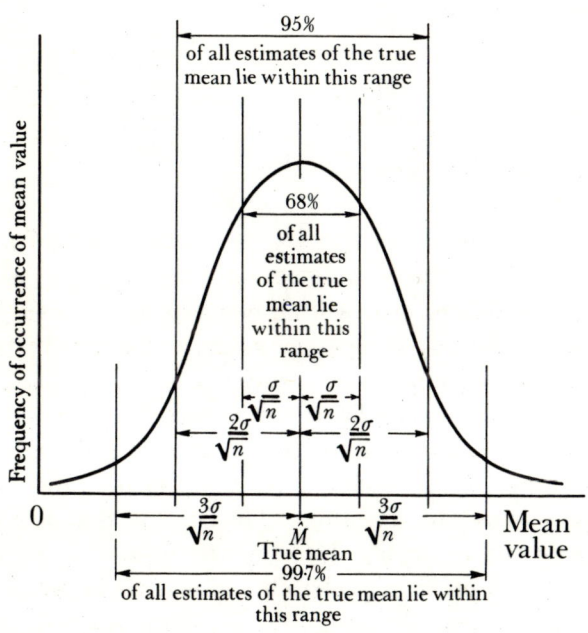

Figure 3.1 Percentage points on the means distribution

So far we have only considered whole number multiples of the standard deviation, i.e., $1\sigma/\sqrt{n}$, $2\sigma/\sqrt{n}$, $3\sigma/\sqrt{n}$, etc., but clearly there will be a percentage figure associated with any multiple, whole number or not, of the standard deviation. Some of these are given in Table 3.1.

Table 3.1

Number of standard deviations	Percentage of values
0·84	60
0·94	65
1·00	68·3
1·04	70
1·15	75
1·28	80
1·44	85
1·50	86·6
1·64	90
1·96	95
2·00	95·4
2·33	98
3·00	99·7

Using Table 3.1 we can say, for example, that 75 per cent of the values obtained will lie between $\hat{M} \pm 1·15\sigma/\sqrt{n}$, 90 per cent will lie between $\hat{M} \pm 1·64\sigma/\sqrt{n}$, and so on, for any percentage figure we choose. We will now go on to see how this knowledge is used in assessing the confidence we can have in estimates.

We have seen that 68 per cent of all values of the mean life obtained from several tests lie within the range true mean $\pm$ standard deviation. This means that for any value of the mean life M_x obtained from a single test there is a 68 per cent probability of it lying in the range $\hat{M} \pm \sigma/\sqrt{n}$, and we can go on to use the other figures: there is a 95 per cent probability that M_x will lie in the range $\hat{M} \pm 2\sigma/\sqrt{n}$, and a 99·7 per cent probability that it will lie in the range $\hat{M} \pm 3\sigma/\sqrt{n}$, and so on. Thus the 'percentage' column given above is in fact a probability figure.

Considering the 68 per cent figure alone, for a moment, we can go on to say that since we are 68 per cent sure of M_x lying in the region $\hat{M} \pm \sigma/\sqrt{n}$, then we are also 68 per cent sure that $\hat{M}$, the true mean, will lie in the region $\hat{M} \pm \sigma/\sqrt{n}$. This is clear from Fig. 3.2.

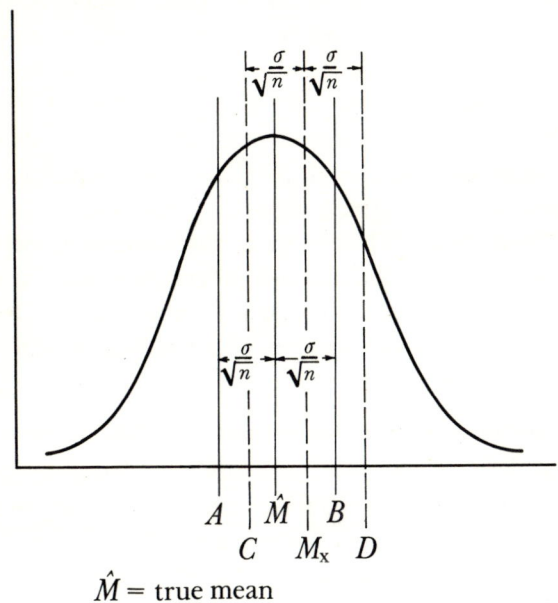

$\hat{M}$ = true mean
M_x = estimate of the true mean

Figure 3.2 Illustration for section 3.2

Clearly if M_x lies between A and B then $\hat{M}$ must lie between C and D since the distance from A to $\hat{M}$ and from $\hat{M}$ to B is the same as the distance from C to M_x and from M_x to D, i.e., *one* standard deviation. So we can now go on to take any one of the estimated means M_x and say that we are

68 per cent sure that the true mean $\hat{M}$ lies

in the range $M_x \pm \sigma/\sqrt{n}$,

95 per cent sure that the true mean $\hat{M}$ lies

in the range $M \pm 2\sigma/\sqrt{n}$,

and so on. In practice, instead of the word 'sure' we say 'confident'. Thus we are 68 per cent confident that $\hat{M}$ lies between $(M_x - \sigma/\sqrt{n})$ and $(M_x + \sigma/\sqrt{n})$. The range $M_x \pm \sigma/\sqrt{n}$ is called the 68 per cent confidence *interval* and the values $M_x - \sigma/\sqrt{n}$ and $M_x + \sigma/\sqrt{n}$ are called the *lower confidence limit* and the *upper confidence limit* respectively for a 68 per cent *level of confidence*. Using the Table 3.1, which we now see gives the number of standard deviations associated with particular percentage levels of *confidence* we can establish confidence limits for any chosen confidence level. Thus from a single wearout test for n failures we obtain the standard deviation σ and the mean wearout life M_x, and we can say with whatever level of confidence we choose between which values the true mean wearout life $\hat{M}$ will lie.

3.3 Confidence limits for exponential distributions

The foregoing discussion of confidence levels and limits applies to normal distributions only. As we have seen, the M T B F, m, for a constant failure rate (either due to chance alone, or chance and wearout if a replacement-on-failure procedure is adopted) is obtained from a failure curve which is of exponential form.

To establish confidence limits for an exponential distribution, the method involving numbers of standard deviations associated with various confidence levels cannot be used directly, since this method applies only to normally distributed variables. Instead, use is made of another well-known statistical distribution called the *chi-square* (pronounced ki-square) distribution. The characteristics and form of this distribution are beyond the scope of this text, but the equations derived from the distribution will be used. The parameter chi, χ, has various values for different values of confidence chosen and for the number of failures occurring in the test from which the figures to be assessed are taken. Values of χ^2 are given in standard statistical tables and are inserted in the following equations.

It can be shown, for a test involving n failures and giving an estimate m of the true MTBF, that for a $100(1 - \alpha)$ per cent confidence level (where α is a constant determined by the required level, i.e., $0\cdot1$ for 90 per cent, $0\cdot2$ for 80 per cent, etc.),

the *upper confidence limit* is $\dfrac{2rm}{\chi^2_{1 - \alpha/2, \, n}}$,

where $\chi_{1 - \alpha/2, \, n}$ is the value of the parameter chi for $1 - \alpha/2$ and for n failures,

and the *lower confidence limit* is $\dfrac{2 \, rm}{\chi^2_{\alpha/2, \, n}}$

where $\chi_{\alpha/2, \, n}$ is the value of the parameter chi for $\alpha/2$ and for n failures. The method of using these equations is clarified in the following example:

Example

An exponentially failing system showed 20 failures in 2 000 component hours† of operation. Determine the confidence limits which the true MTBF will lie between for a 98 per cent level of confidence.

An estimate of the MTBF is $2\ 000/20 = 100$ hours.

The level of confidence is 98 per cent. Thus $100(1 - \alpha) = 98$, and therefore $\alpha = 0\cdot02$.

† That is, total survival hours.

Thus $\alpha/2$ is $0 \cdot 01$ and $1 - \alpha/2$ is $0 \cdot 99$. Hence, using Table 3.2† for 20 failures and a value of $\alpha/2$ equal to $0 \cdot 01$ χ^2 is $63 \cdot 7$; for 20 failures and a value of $1 - \alpha/2$ equal to $0 \cdot 99$, χ^2 is $22 \cdot 16$.

Table 3.2 Values of χ^2

Values of $\alpha/2$ or $(1 - \alpha/2)$

Failures	0·995	0·99	0·975	0·95	0·05	0·025	0·01	0·005
5	2·16	2·56	3·25	3·94	18·3	20·48	23·2	25·19
10	7·43	8·26	9·59	10·85	31·41	34·17	37·57	39·99
15	13·79	14·95	16·79	18·49	43·77	46·98	50·89	53·67
20	20·71	22·16	24·43	26·51	55·76	59·34	63·7	66·8
25	27·99	29·71	32·36	34·76	67·5	71·42	76·2	79·5
30	35·53	37·48	40·48	43·19	79·08	83·30	88·4	92
35	43·28	45·44	48·76	51·74	90·43	95·02	100·4	104·2
40	51·17	53·54	57·15	60·39	101·9	106·6	112·3	116·3

Thus, using the equations already given,

$$\text{lower confidence limit} = \frac{2\,nm}{\chi^2_{\alpha/2,\,n}} = \frac{40 \times 100}{63 \cdot 7}$$

$$= 62 \cdot 8 \text{ hours}$$

$$\text{upper confidence limit} = \frac{2\,nm}{\chi^2_{1-\alpha/2,\,n}} = \frac{40 \times 100}{22 \cdot 16}$$

$$= 181 \text{ hours}$$

so we can say that we are 95 per cent confident that the true MTBF lies within the range $62 \cdot 8$ to 181 hours.

† A selection of values based on 'Table of percentage points of the χ^2 distribution', Catherine M. Thompson, *Biometrika*, vol. 32 (1941).

4. System reliability

4.1 Inclusion of subunits in systems: intermittent operation

A system very often contains subunits which are not required to function the whole time during the system operating period. In these cases care must be taken when assessing the MTBF, failure rate, and reliability to take into account the unit operating time and the fact that it is not equal to the system operating time.

We shall consider the exponential case only since this is the most commonly found. Consider a component or subunit having a reliability R_c for t_c component operating hours. R_c is given by

$$R_c = e^{-t_c/m_c}$$

where m_c is the mean time between failure of the component expressed in component operating hours. If now this component is inserted in an exponentially failing system which is operating for t_s hours, of which the component is to operate only t_c hours, then the component reliability for t_s system operating hours will be the same as for t_c component operating hours. The reliability of the component, expressed as a function of system hours, is given by

$$e^{-t_s/m_s},$$

where m_s is the mean time between failure of the component expressed in system operating hours.

The reliability of the component is the same however it is expressed, since it is operated for t_c hours whether alone or as a part of a system. Thus

$$R_c = e^{-t_c/m_c} = e^{-t_s/m_s}$$

and therefore

$$\frac{t_c}{m_c} = \frac{t_s}{m_s}$$

or

$$\frac{m_c}{m_s} = \frac{t_c}{t_s}. \qquad (4.1)$$

The ratio t_c/t_s expresses the fraction of system time during which the component is required to operate and is called the *duty cycle* of the component, indicated by d. So we have

$$m_s = \frac{m_c}{d} \quad \text{where} \quad d = \frac{t_c}{t_s}. \qquad (4.2)$$

Illustrating this numerically, suppose we have a component operating for 5 hours in a system time of 25 hours and having a reliability of 0·9. MTBF expressed in component hours, m_c, is given by

$$0·9 = e^{-5/m_c},$$

so that

$$e^{5/m_c} = 1/0·9.$$

Thus

$$\frac{5}{m_c} \ln e = \ln\left(\frac{1}{0·9}\right),$$

and

$$m_c = 5/\ln(1/0·9)$$
$$= 47·5 \text{ hours.}$$

Thus we may expect on average a component failure for every 47·5 hours of its operation.

The duty cycle $= \frac{5}{25} = \frac{1}{5}$, so that the MTBF expressed in system operating hours, m_s, is given by

$$m_s = 5 \times 47·5$$
$$= 237·5 \text{ hours.}$$

On average a component fails for every 237·5 hours of operation of the system. This is quite logical since for 237·5 hours of system operation the component operates only $\frac{1}{5}$ of this, i.e., 43·5 hours, its MTBF. The above assumes zero failure probability as the component is switched into and out of use. If there is such a probability this must, of course, be taken into account when assessing the system MTBF.

4.2 System reliability

In calculating the reliability of a system made up of a number of components or individual complete units, each having their own reliability, the *type* of system must first be determined. There are two types—series systems and parallel systems.

A series system is one in which failure of one of the sub-units or components means failure of the system as a whole. A parallel system is one which does not fail until all subunits or components have failed.

It will be remembered that reliability is the probability of survival; in computing system reliability various well-established laws of the mathematics of probability are used. The relevant laws are discussed below.

4.3 Laws of probability relevant to reliability calculations

1. If X and Y are two independent events and P_x is the probability that X will occur and P_y is the probability that Y will occur, then the probability that *both* events X and Y will occur, P_{xy}, is given by

$$P_{xy} = P_x P_y \qquad (4.3)$$

2. If the two events can occur simultaneously, the probability that either X or Y or both X and Y will occur, P_{x+y}, is given by

$$P_{x+y} = P_x + P_y - P_x P_y \qquad (4.4)$$

Both of these laws may be extended to cover more than two events.

4.4 Reliability of a series system

The reliability of a series system or probability of survival of the system is the probability of *all* the components surviving since a failure of only one component means overall system failure.

If R_s is the system reliability and R_1, R_2, etc., are the reliabilities of the system components for the same time period, then using law 1:

$$R_s = R_1 R_2 \qquad (4.5)$$

In this case the 'event' described in law 1 is survival. This expression is called the *product law of reliabilities* for series systems. For a system having more than two components the additional reliabilities, R_3, R_4, etc., are included in the product term.

4.5 Unreliability of a series system

The unreliability of a series system or probability of failure of the system is the probability of at least *one* of the system components failing.

If Q_s is the system unreliability and Q_1, Q_2, etc. are the unreliabilities of the system components for the same time period, then using law 2:

$$Q_s = Q_1 + Q_2 - Q_1 Q_2$$
$$\text{for a system having two components} \qquad (4.6)$$

It will be recalled that unreliability = 1 − reliability, i.e., in general $Q = 1 - R$ so that

$$Q_1 = 1 - R_1,$$
$$Q_2 = 1 - R_2,$$

and substituting in the equation for Q_s we see that

$$Q_s = (1 - R_1) + (1 - R_2) - (1 - R_1)(1 - R_2)$$
$$= 1 - R_1 + 1 - R_2 - 1 + R_1 + R_2 - R_1 R_2$$
$$= 1 - R_1 R_2$$
$$= 1 - R_s \qquad (4.7)$$

which is to be expected. In this case the 'event' described in law 2 is failure.

The unreliability expression for a system having more

29

than two components is more complex than that above but using $Q_s = 1 - R_s$ the reliability may first be determined using eq. (4.5) and then the unreliability determined using eq. (4.7).

4.6 Reliability of a parallel system

The reliability of a parallel system, or probability of survival of the system, is the probability of at least *one* component surviving since, provided that at least one component survives in a parallel system, the system will not fail.

If R_p is the system reliability and R_1, R_2 are the component reliabilities, then, using law 2,

$$R_p = R_1 + R_2 - R_1 R_2. \qquad (4.8)$$

In this case, the 'event' described in law 2 is survival. For a system having more than two components, the reliability expression is more complex but, as we shall see, an easy method of calculation is to determine the unreliability first and use eq. (4.10) given below.

4.7 Unreliability of a parallel system

The unreliability of a parallel system, or probability of system failure, is the probability of *all* components failing since, for a parallel system, even if only one component survives the system has not failed.

If Q_p is the system unreliability, and Q_1, Q_2 the component unreliabilities, then, using law 1,

$$Q_p = Q_1 Q_2. \qquad (4.9)$$

In this case, the 'event' described in law 1 is failure. This expression is called the *product law of unreliabilities* for parallel systems. For a system having more than two components the additional unreliabilities are included in the product term.

Since $Q_1 = 1 - R_1$ and $Q_2 = 1 - R_2$,

$$
\begin{aligned}
Q_p &= (1 - R_1)(1 - R_2) \\
&= 1 - R_1 - R_2 + R_1 R_2 \\
&= 1 - (R_1 + R_2 - R_1 R_2) \\
&= 1 - R_p,
\end{aligned}
\tag{4.10}
$$

which is to be expected. This equation is most useful for determining the system reliability R_p after having found the system unreliability Q_p. It is easier to compute Q_p than R_p in the first instance, because Q_p is contained in the product law of unreliabilities while the expression for R_p becomes increasingly complex as the number of subunits is increased.

4.8 Systems containing exponentially failing units

Whether or not a system made up of units which are failing exponentially behaves overall in an exponential fashion depends on the system type. It is found that series systems do behave exponentially and their reliability may be expressed in the familiar $e^{-\lambda t}$ form, whereas parallel systems do not and the form of their reliability equation depends upon the number of subunits.

Series systems
The reliability equation as shown above is

$$
R_s = R_1 R_2 R_3 \cdots
$$

where R_s is the system reliability and R_1, R_2, R_3, etc., are the subunit reliabilities.

If

$$
R_1 = e^{-\lambda_1 t}, \quad R_2 = e^{-\lambda_2 t}, \quad R_3 = e^{-\lambda_3 t}, \quad \text{etc.},
$$

where λ_1, λ_2, λ_3 are the respective failure rates of the subunits, then

$$
\begin{aligned}
R_s &= e^{-\lambda_1 t} \, e^{-\lambda_2 t} \, e^{-\lambda_3 t} \\
&= e^{-(\lambda_1 + \lambda_2 + \lambda_3)t}
\end{aligned}
\tag{4.11}
$$

which is, of course, of general exponential form, the system failure rate being the *sum* of the individual failure rates. The system failure rate clearly increases as the number of subunits is increased. Since the overall behaviour is exponential, i.e., we can express the reliability R_s in the form

$$R_s = e^{-\lambda_s t},$$

where λ_s, the system failure rate, is given by

$$\lambda_s = \lambda_1 + \lambda_2 + \lambda_3. \qquad (4.12)$$

Then the MTBF for a series system, m_s, is equal to the reciprocal of the system failure rate:

$$m_s = \frac{1}{\lambda_s} = \frac{1}{\lambda_1 + \lambda_2 + \lambda_3}. \qquad (4.13)$$

This reciprocal equation applies only to components or systems whose reliability is expressible in exponential form, i.e., having constant failure rate.

For a series system containing n similar units of equal reliability the

$$\text{system reliability} \quad R_s = e^{-n\lambda t} \quad (4.14)$$

$$\text{system failure rate} \quad \lambda_s = n\lambda \quad (4.15)$$

$$\text{system MTBF} \quad m_s = \frac{1}{n\lambda} \quad (4.16)$$

Parallel systems
As was stated earlier, the form of the reliability expression for a parallel system depends upon the number of subunits. For a simple two-unit system

$$R_p = R_1 + R_2 - R_1 R_2; \qquad (4.17)$$

for a three-unit system

$$R_p = R_1 + R_2 + R_3 - R_1 R_2 \\ - R_2 R_3 - R_1 R_3 + R_1 R_2 R_3, \qquad (4.18)$$

and the expression becomes more complex as the number of units is increased.

If we write $R_1 = e^{-\lambda_2 t}$, $R_2 = e^{-\lambda_2 t}$, etc., as we did for series systems, then the system reliability for a two-unit system is given by:

$$R_p = e^{-\lambda_1 t} + e^{-\lambda_2 t} - e^{-(\lambda_1 + \lambda_2)t} \qquad (4.19)$$

for a three-unit system by

$$\begin{aligned} R_p = {} & e^{-\lambda_1 t} + e^{-\lambda_2 t} + e^{-\lambda_3 t} \\ & + e^{-(\lambda_1 + \lambda_2)t} + e^{-(\lambda_2 + \lambda_3)t} \\ & + e^{-(\lambda_1 + \lambda_3)t} - e^{-(\lambda_1 + \lambda_2 + \lambda_3)t}. \end{aligned} \qquad (4.20)$$

Clearly, these equations are not of simple exponential form, and we cannot express the overall system reliability in the form $e^{-\lambda_p t}$ where λ_p is a constant system failure rate. The MTBF may still be obtained by integration of the reliability expression, as was done in the series case, since, as was shown earlier, this method of finding the MTBF does not rely on the expression being of exponential form. However, for a parallel system, the MTBF is not the reciprocal of the system failure rate, but depends upon the number of subunits.

It can be shown that the MTBF, m_p, for a two-unit system, is given by

$$m_p = \frac{1}{\lambda_1} + \frac{1}{\lambda_2} - \frac{1}{\lambda_1 + \lambda_2}; \qquad (4.21)$$

for a three-unit system it is given by

$$\begin{aligned} m_p = {} & \frac{1}{\lambda_1} + \frac{1}{\lambda_2} + \frac{1}{\lambda_3} - \frac{1}{\lambda_1 + \lambda_2} - \frac{1}{\lambda_2 + \lambda_3} \\ & - \frac{1}{\lambda_1 + \lambda_3} + \frac{1}{\lambda_1 + \lambda_2 + \lambda_3}, \end{aligned} \qquad (4.22)$$

where λ_1, λ_2, λ_3, are the unit failure rates respectively, and for an n-unit system, each unit having the same

33

failure rate λ

$$m_{\text{p}} = \frac{1}{\lambda} + \frac{1}{2\lambda} + \frac{1}{3\lambda} + \cdots + \frac{1}{n\lambda}. \qquad (4.23)$$

The system failure rate for a parallel system is not constant but is time dependent.

Series and parallel systems
—reliability compared

In both system types the overall reliability is dependent upon the number of subunits. To illustrate the effect of the series and parallel connection of the system subunits, consider again the reliability expressions. For a two-unit system having unit reliabilities R_1, R_2

$$\text{series system reliability} = R_1 R_2$$
$$\text{parallel system reliability} = R_1 + R_2 - R_1 R_2.$$

Bearing in mind that reliability can never exceed unity, examination of the two expressions shows that

$$R_1 + R_2 - R_1 R_2 \geqslant R_1 R_2,$$

the two being equal only when $R_1 = R_2 = 1$ and at all other values of R_1 and R_2 the left-hand side being greater than the right-hand side. This means then, that the reliability for a two-unit parallel system is greater than that for a two-unit series system except when the subunits have equal reliabilities equal to unity. In this case series and parallel systems will have equal reliabilities. A similar observation may be made concerning reliabilities of three-unit systems:

$$\text{series system reliability} = R_1 R_2 R_3$$
$$\begin{aligned}\text{parallel system reliability} = {}& R_1 + R_2 + R_3 \\ & - R_1 R_2 - R_2 R_3 - R_1 R_3 \\ & + R_1 R_2 R_3.\end{aligned}$$

Clearly the first six terms of the parallel reliability expression must have an overall value of zero or greater than zero since R_1, R_2, or R_3 must all be unity or less than unity. The seventh term is, in fact, the series reliability so we have

(parallel reliability) = (terms $\geqslant 0$) + (series reliability),

which is obviously greater than or equal to the series reliability. It can be shown that for any number of units the parallel reliability is greater than or equal to the series reliability. Physically, this is logical since for a parallel system all units except one can fail before system failure, whereas, for a series system, only one unit needs to fail for system failure. Typical sets of reliability curves for one-, two-, and three-unit systems in both connection modes are shown in Fig. 4.1.

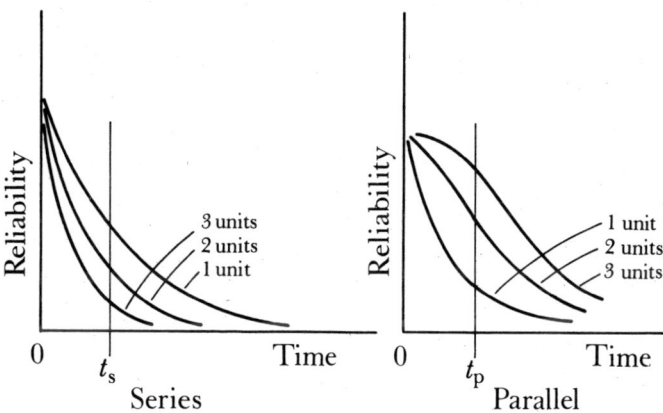

Figure 4.1 Reliability curves for series and parallel systems

Notice that for a series system the reliability at any time t_s is reduced as the number of subunits is increased. The reliability curve is always exponential regardless of the number of subunits. For a parallel system the reliability at

35

any time t_p is increased as the number of subunits is increased. For systems containing more than one unit, the reliability curve ceases to be exponential, as was explained earlier.

This improved reliability for parallel-connected systems is used in schemes to reduce risk of failure, such as stand-by systems, in which several equal units are allowed to stand idle ready to take over should the operational unit fail. Such schemes are said to employ *parallel redundancy*.

5. System maintenance and inspection

5.1 Effects of inspection and overhaul on reliability

Our study of reliability so far has been concerned with components or systems which are not subject to regular inspection and overhaul. It is logical that reliability may be improved if regular maintenance is carried out, and in this section we shall examine briefly how overhaul schedules affect the MTBF, percentage failure rate, and thus the reliability.

In calculating overhaul schedules, the cost of part replacement and labour must be balanced against any improvement in reliability which is obtained as the period between scheduled overhauls, i.e., the operating period, is prolonged. The reliability improvement obtained by planned maintenance depends upon the type of system, whether series or parallel, and the type of failure, whether chance or wearout, exponential or otherwise, which is anticipated.

Wearout failure may be reduced or even avoided if the operating period is kept to a value below that at which wearout takes effect. In practice, this means keeping the operating period shorter than $(M - 3\sigma)$, where M is the mean wearout life and σ the standard deviation of the normal wearout distribution, for single components or shorter than $(M - 5\sigma)$ or even $(M - 6\sigma)$ for multicomponent systems.

The reduction in chance failures depends to a considerable degree on system connection.

A parallel system requires failure of all its subunits before system failure occurs and it may happen that several of these may have failed due to chance, even though the system still functions. Regular overhaul will reveal these chance-failed subunits, which may then be replaced, thus reducing the risk of overall system failure due to chance alone. In a series system only one subunit need fail by chance to cause chance failure of the whole system. Regular overhaul cannot reveal subunits which are about to fail by chance and so the overall risk of system chance failure is unaffected by overhaul schedules.

It is found in practice that the series system MTBF remains constant when overhauls are employed, and the main reason for their use is to reduce the risk of wearout failure. For a parallel system the system MTBF depends upon the time between overhauls to a considerable degree

and planned maintenance can reduce the risk both of
wearout and chance failure.

5.2 Maintainability, availability, and dependability

The probability of performing a given repair on a unit or
system within a given time is called the maintainability of
the unit or system. It is dependent on the type of unit or
system to be repaired, on its location, and on the efficiency
and skill of the repair crew. It is important that maintainability is kept as high as possible if optimum use is to be
derived from the system or unit.

The utilization factor of a unit or system is defined as
the ratio between the operating time and the total time
required to ensure operation. The total time is the sum of
the maintenance time, including scheduled and unscheduled (due to failure) maintenance, idle time, which
may occur between completion of maintenance, and use
due to administrative difficulties, etc., and actual operating time. Mathematically

$$U = \frac{t_{op}}{t_{main} + t_{idle} + t_{op}}, \qquad (5.1)$$

where U is utilization factor, t_{op} is operating time, t_{main} is
maintenance time, t_{idle} is idle time. The utilization factor
is a maximum when t_{idle} is zero and t_{main} is as small as
possible. Maximum utilization factor is called the availability of the system or unit.

Availability

$$A = U_{max} = \frac{t_{op}}{t_{main (min)} + t_{op}}, \qquad (5.2)$$

where U_{max} is maximum utilization factor, t_{op} is operating
time, $t_{main (min)}$ is maintenance time minimum value. The
dependability of a system or unit is defined as the availability when the maintenance time includes only that

39

used for unscheduled overhauls necessitated by failure, i.e., when time allocated to scheduled overhauls, for preventive maintenance, is omitted from the calculation. Clearly, for the best use of an equipment, its availability and dependability figures must be as high as possible. The factors affecting these figures are numerous and include good design, high component reliability, ease of maintenance and repair, skill and efficiency of personnel and so on. The fields of study involved include material engineering, design engineering, reliability engineering, and organization and methods engineering. A more detailed study will be found in more advanced texts on these subjects.

6. Causes of failure— component stress

The stresses acting on components within a system may be divided into two categories. These are:

1. Environmental stresses, characterized by their existence under operating or nonoperating conditions.

2. Operating stresses, characterized by their existence only under operating conditions.

6.1 Environmental stresses

Environmental stresses contribute significantly to the reduction of the component mean life and thus to wearout

4*

failure. Their contribution to chance failure may be considered slight. Operating stresses, on the other hand, contribute to both chance and wearout failure.

Environmental stresses may be further subdivided by cause, as follows:

Temperature

The ambient temperature of the environment in which a component is situated is of the utmost importance. Bearing in mind that the working temperature will probably be higher due to operating, as discussed below, it is essential that a component is operated well within the rated extremes indicated by the manufacturer. The intended geographical area of use is important since ambient temperature fluctuations vary considerably throughout the world. For example, the desert regions experience an approximate range of $-15°C$ to $65°C$, tropical regions a range of $20°C$ to $45°C$, maritime regions a range of $-10°C$ to $28°C$, polar regions a range of $-40°C$ to $35°C$, and the atmosphere, a region of increasing importance, may vary from the surface temperature to $-40°C$ in the troposphere (up to about 12 km from ground level), $35°C$ in the stratosphere (about 37 km from ground level), down again to $-80°C$ in the mesophere (between 40 km and 75 km above ground level). Equipment situated in mobile systems may have to experience a range anywhere between these limits, and temperature considerations of components in systems such as aircraft and space vehicles are detailed and of paramount importance. A useful approximation in these considerations is the Arrhenius law which indicates that the failure rate doubles with every $10°C$ rise in temperature. This is considered in a little more detail under 'derating'.

Humidity

The content of water vapour in the air is an important

42

factor in reliability design. Again, geography indicates wide variations of relative humidity in various climatic conditions ranging from about 3 per cent in desert regions to 100 per cent in tropical locations. Upper atmosphere humidity is low, being about 2 per cent in the troposphere and falling as the height above ground increases. To a certain extent the effects of increased humidity are somewhat compensated by increasing temperature drying out the component during use.

Atmospheric pressure

External pressure on a component or system due to the atmosphere varies relatively slightly at surface level, but in airborne systems, especially in space vehicles, pressure extremes may vary considerably. At ground level, normal pressure is taken as equivalent to 76 cm of mercury, but by the time the troposphere is reached atmospheric pressure has fallen to about 25 cm Hg and above the stratosphere is down to a few millimetres of mercury.

Chemical content

As might be expected, the content of the surrounding air is important. Chemical reaction, corrosion, or formation of adverse compounds may hasten failure. Again, it should be noted that airborne systems need most thought since it is known that the atmosphere varies from a nitrogen–oxygen mixture, containing only about 0·01 per cent hydrogen, at low altitude, to a 95 per cent hydrogen content mixture, containing virtually no nitrogen or oxygen, at altitudes in the troposphere and above. Ground level atmosphere may also be contaminated by chemical impurity, depending on the situation of the system, and special-case designs are essential for specific-purpose systems. Impurities in the air which can be considered loosely under this heading also include the heavy dust density in desert regions and the excess salinity of the atmospheric water vapour in maritime coastal regions.

Radiation

Electromagnetic radiation, especially of a nuclear nature, can adversely affect correct functioning of certain types of component, especially bipolar solid state devices. (Unipolar devices are somewhat more resistant.) Earthbound systems controlling or measuring nuclear energy processes are particularly vulnerable since, of necessity, they are situated in regions where humans are not, and again, airborne and spaceborne vehicles must withstand increasing bombardment as the earth is left behind. Considerable study of failure causation due to excessive radiation in space systems is being made at present in the U.S.A. and the U.S.S.R.

6.2 Operating stresses

Operating stresses may be further subdivided by cause, as follows:

Voltage

The stresses due to applied voltage and the subsequent internal electric field set up within the molecular structure of electrical components are self evident. An important part of the reliability consideration is not so much the steady-state stresses but those due to transient surges on switching, which may exceed normal operational value by many times. It is, of course, essential to maintain the applied voltage at, or even below, the rated value, as discussed below. Components utilizing a static electric field, such as capacitors, have been shown to behave according to a fifth power law, which indicates that component failure rate is approximately proportional to the fifth power of the applied potential difference. This is discussed under 'derating' below.

Current and frequency

Reliability considerations for component current and frequency contribution to failure are similar to those of voltage. Again, spasmodic fluctuation in excess of quiescent values must be anticipated, but compensation may be affected to some extent by the use of derating procedures described below.

A third group of stresses of equal if not greater importance cannot be categorized as the above. These are mechanical stresses due to shock, vibration, friction, acceleration, etc., which contribute to both chance and wearout failure. They may be caused environmentally or by the normal functioning of the component. Again, physical location is particularly important and, in general, failure rate in mobile systems is higher than in static systems. Mechanical stress in systems used in space vehicles or other devices utilizing rocket propulsion is enormous and, as is so often the case in other fields of study, here, where failure is most likely it is least tolerable.

6.3 Stress reduction and derating

The effects of environmental stresses due to humidity, atmospheric pressure, radiation, and chemical content of the surroundings may be minimized to some extent by encapsulation techniques, although no form of encapsulation completely protects a component. Encapsulation techniques range from provision of weatherproof, dustproof, separate covers such as those used in discrete component manufacture (transistor and diode opaque containers sealed at the base) to solid encapsulation (potting) in resin compounds. Potting also assists in the protection against shock and vibration effects. The form of casting resin must be carefully chosen since, on setting, various mechanical stresses are set up which may damage the potted component.

Reliability of electronic systems has been considerably improved by the use of printed circuit boards encapsulated in a rubber compound. The compound offers protection against environmental effects as listed, and, in addition, against mechanical stresses induced by shock and vibration. If necessary, it can be removed by slicing for maintenance of individual components. Even better reliability figures are obtainable from integrated circuits in which complete circuits or even whole systems are contained within a single chip of silicon. Maintenance on individual components is, of course, not possible, and the whole unit is replaced when necessary, but modern production techniques are constantly reducing unit costs, and it is estimated that complete circuits may soon cost less than a single discrete component does at present.

A technique of solderless jointing, developed by Bell Laboratories, is presently achieving reduced failure rates; the wire is wrapped under pressure around terminal posts, which are square or rectangular in cross section, and the wire 'bites' into the corners of the post. The possibility of 'dry' soldered joints is thus eliminated.

Mechanical stresses due to shock or vibration may be reduced by correct mounting on resilient supports having a natural frequency of vibration either greater than or less than that which may be expected, depending on the purpose of the component or system.

Operational stresses may be reduced by a process known as derating, in which the component is operated under conditions of voltage, current, frequency, etc., which are below the rated values. To determine the optimum conditions for voltage or temperature, sets of failure-rate curves are obtained for the particular component in the following manner. Bearing in mind the

Arrhenius law and the fifth power law quoted above, it is assumed that the failure rate for any particular set of conditions, λ_x, is related to the failure rate under measured conditions, λ_m, by the formula

$$\lambda_x = \lambda_m \left(\frac{V_x}{V_m}\right)^n K^{(t_x - t_m)}, \tag{6.1}$$

where V_m, V_x denote the voltage at the measured condition and at any other condition respectively, t_m, t_x denote the temperature at the measured condition and at any other condition respectively, n is a constant (assumed to be 5 in the fifth power law) for a particular component, K is a constant showing the effect of temperature for a particular component. The constants n and K may be determined by tests involving firstly keeping the voltage constant, and thus determining K, and secondly keeping temperature constant, and thus determining n. Mechanical stresses are maintained constant throughout. This may be clarified by considering eq. (6.1), firstly for constant voltage, giving

$$\lambda_x = \lambda_m K^{(t_x - t_m)}$$

and hence

$$K = \left(\frac{\lambda_x}{\lambda_m}\right)^{1/(t_x - t_m)}, \tag{6.2}$$

and secondly for constant temperature, giving

$$\lambda_x = \lambda_m \left(\frac{V_m}{V_x}\right)^n$$

and hence

$$n = \frac{\ln \lambda_x/\lambda_m}{\ln V_m/V_x} \tag{6.3}$$

Hence, knowing n and K, curves may be plotted for a number of values of voltage or temperature without further measurement. Care is necessary, however, since n and K may be assumed constant only over limited ranges.

For current or frequency more detailed measurements are necessary since the same approximation cannot be made. However, similar failure rate curves may be determined and optimum conditions chosen.

The optimum conditions are determined by an examination of all failure rate curves for the various variable quantities and any additional relationships connecting these variables (for example, a relationship between current or voltage and additional temperature rise due to operation). It is found that a dramatic reduction in failure rates may be accomplished by operation under conditions where voltage temperature, etc., are between one-half and one-third of the rated values.

Worked examples

A summary of equations appears on p. 58; references in these examples are to this set.

1. In a test to determine the MTBF of a certain component, 100 were tested for a period of 4 000 hours. The times to failure of the components are as shown in Table 1. Assuming that wearout failure can be ignored, calculate the failure rate and the MTBF.

Table 1

Number of components	Time to failure (hours)
1	250
1	300
4	415
5	800
3	1 200
86	No failure

Total survival hours
$$= 250 + 300 + 4 \times 415 + 5 \times 800$$
$$+ 3 \times 1\,200 + 86 \times 4\,000$$
$$= 550 + 1\,660 + 4\,000 + 3\,600 + 344\,000$$
$$= 353\,810 \text{ hours.}$$

Total failures $= 14$.

Using eq. (11),

$$\text{M T B F} = \frac{353\,810}{14}$$

$$= 25\,270 \text{ hours}$$

Since we can ignore wearout failure we can assume an exponential relationship, and hence that the failure rate is the reciprocal of the M T B F.

Thus

$$\lambda = \frac{1}{25\,270} = 3{\cdot}957 \times 10^{-5}/\text{hour}$$

$$= 3{\cdot}957 \text{ per cent}/1\,000 \text{ hours.}$$

2. Calculate the failure rate of a component having a reliability figure of $0{\cdot}91$ for a period of 400 hours. Assume exponential failure. Calculate also the unreliability for 800 hours.

Applying eq. (1)

$$0{\cdot}91 = e^{-400\lambda},$$

where λ is the failure rate. Rearranging,

$$1/e^{400\lambda} = 0.91$$

i.e.

$$e^{400\lambda} = 1/0.91$$

Taking natural logarithms,

$$400\lambda = \ln\left(\frac{1}{0.91}\right).$$

Therefore,

$$\lambda = \frac{1}{400}\ln\left(\frac{1}{0.91}\right)$$
$$= 0.000\ 237$$
$$= 23.7 \text{ per cent}/1\ 000 \text{ hours.}$$

The unreliability could be calculated using eq. (3). However, an easier method is to calculate the reliability first, with the method shown below, and then apply eq. (2). We know that

$$0.91 = e^{-400\lambda},$$

and since

$$e^{-800\lambda} = (e^{-400\lambda})^2$$

then

$$e^{-800\lambda} = 0.91^2$$
$$= 0.8281$$

which is the reliability for 800 hours.

By eq. (2), the unreliability for 800 hours equals $1 - 0.8281 = 0.1719$. This means that we could reasonably expect 17.19 per cent of the components to fail during the 800-hour period.

3. Calculate the number of components having the failure rate of question (2) above which are still alive after

a period of 800 hours if at the start of the period there are 1 000.

From eq. (4)

$$N_s = 1\ 000 \times e^{-0.000\ 237 \times 800}$$
$$= 1\ 000 \times 0.828\ 1$$
$$= 828$$

4. A wearout test gave the life data shown in Table 2. Calculate the upper and lower confidence limits for the true mean wearout life at (a) 68 per cent confidence level, (b) 85 per cent confidence level.

Table 2

Number of components	Life
5	390
8	450
10	500
7	550
6	600

Sum of lives is $5 \times 390 + 8 \times 450 + 10 \times 500 + 7 \times 550 + 6 \times 600$

$$= 1\ 950 + 3\ 600 + 5\ 000 + 3\ 850 + 3\ 600$$
$$= 18\ 000 \text{ hours.}$$

Total number of components is 36. Therefore, estimated Mean wearout life

$$M_x = 18\ 000/36 = 500 \text{ hours.}$$

The differences and squares of differences between the mean and actual lives are as shown in Table 3.

The standard deviation, from eq. (12) is

$$\sqrt{\frac{5 \times 12\ 100 + 8 \times 2\ 500 + 7 \times 2\ 500 + 6 \times 10\ 000}{36}}$$
$$= 73.7 \text{ hours.}$$

Table 3

Number	Life	Difference	Square
5	390	110	12 100
8	450	50	2 500
10	500	—	—
7	550	50	2 500
6	600	100	10 000

Notice that the calculation overall has been reduced by the fact that we are told 5 components expired after 390 hours, 8 components expired after 450 hours, and so on. It is unlikely in practice that this would be so, and in fact all 36 components would probably have different lives. In a case such as that the lives are grouped in small *class intervals*, and the mean life of each group determined. Thus, in the above example, the figure 390 is already a mean life for the 5 components to which it refers; the 5 components in the original test probably had different lives spread about the mean 390 hours.

Using the expressions given earlier for confidence intervals for a normal distribution, we can say that the *true* mean wearout life M will lie between

$$\left(500 \pm \frac{73 \cdot 7}{\sqrt{36}}\right) \text{hours at a 68 per cent}$$
confidence level,

$$\left(500 \pm 1 \cdot 44 \times \frac{73 \cdot 7}{\sqrt{36}}\right) \text{hours at an 85 per cent}$$
confidence level.

(The figure of 1·44 is obtained from the table of standard deviations and percentage of values given earlier.) That is,

$(500 \pm 12 \cdot 28)$ hours at a 68 per cent confidence level,

$(500 \pm 17 \cdot 77)$ hours at an 85 per cent confidence level.

Thus, we are 68 per cent confident that the true mean life lies between 512·28 hours and 487·72 hours and 85 per cent confident that it lies between 517·77 hours and 482·23 hours.

5. A radio receiver has the failure rates shown in Table 4 for its subunits. Calculate the reliability over a period of 5 000 hours operation.

Table 4

Unit	Number	Failure rate (per cent 1 000 hours)
IF amplifier	4	0·1
AF amplifier	3	0·04
Oscillator	1	0·08
Power supply	1	0·9
Power output	1	1·2

A radio receiver may be considered a series system, since a failure of one of the subunits leads to zero output, i.e., overall failure. Thus, we can apply eqs. (4.11) and (4.12).

Total failure rate

$$= 4 \times 0·1 + 3 \times 0·04 + 0·08 + 0·9 + 1·2$$
$$= 0·4 + 0·12 + 0·08 + 0·9 + 1·2$$
$$= 2·7 \text{ (per cent/1 000 hours)}$$

and the reliability R_s is given by

$$R_s = e^{-2·7 \times 500 \times 10^{-5}}$$
$$= 0·837$$
$$R_s = 83·7 \text{ per cent.}$$

Notice that the failure rate expressed as a percentage per 1 000 hours is converted before insertion into the reliability equation.

6. Compare the reliability of a series system with a parallel system, when each contain 3 subunits having reliabilities 0·9, 0·8, and 0·7 respectively.

From eq. (20),

> series system reliability $= 0.9 \times 0.8 \times 0.7$
> $= 0.504.$

From eq. (33),

parallel system reliability
$= 0.9 + 0.8 + 0.7 - 0.9 \times 0.8$
$\quad - 0.9 \times 0.7 - 0.8 \times 0.7 + 0.9 \times 0.8 \times 0.7$
$= 2.4 - 0.72 - 0.63 - 0.56 + 0.504$
$= 0.994.$

7. A subunit having a failure rate of 10 per cent/1 000 hours is included in a system which is to operate for 2 000 hours. The component duty cycle is 0.5. Calculate the component reliability and the component MTBF in system hours assuming exponential failure.

From eq. (16),

> component operating hours $= 0.5 \times 2\,000$
> $= 1\,000$ hours.

From eq. (1),

> reliability of the component $= e^{-10 \times 1\,000 \times 10^{-5}}$
> $= 0.904\,8.$

Assuming exponential failure, component MTBF from eq. (6)

$$= \frac{1}{10 \times 10^{-5}}$$
$$= 10\,000 \text{ hours.}$$

In system hours, the component MTBF from eq. (17)

$$= \frac{10\,000}{0.25}$$
$$= 40\,000 \text{ hours.}$$

55

8. Over a period of 105 hours, out of a total of 100 components, one failed at each of the times shown below:

9, 10, 20, 30, 50, 70, 80, 80, 100, 101 (hours).

Calculate the confidence limits for the failure rate of these components at a 90 per cent confidence level. Wearout failure may be neglected.

Total survival hours of failed components

$$= 9 + 10 + 20 + 30 + 50 + 70 + 160 + 100 + 101$$
$$= 550 \text{ hours.}$$

Total survival time of non-failed components

$$= 90 \times 105$$
$$= 9\ 450 \text{ hours.}$$

Thus

$$\text{Total survival hours} = 9\ 450 + 550$$
$$= 10\ 000 \text{ hours.}$$

Hence an estimate of the mean life is given by

$$10\ 000/10 = 1\ 000 \text{ hours.}$$

Since wearout failure is to be neglected, an exponential failure rate may be assumed, indicating the use of the chi-square distribution. For a 90 per cent confidence level $1 - \alpha = 0.9$, whence $\alpha/2 = 0.05$ and $1 - \alpha/2 = 0.95$. For 10 failures the value of χ^2 is 10.85 and 31.41 (see table 3.2), and so, since the estimated mean time between failures is 1 000 hours, the upper confidence limit is

$$20\ 000/10.85 = 1\ 844 \text{ hours,}$$

and the lower confidence limit is

$$20\ 000/31.41 = 637 \text{ hours.}$$

Since the exponential distribution is assumed, the percentage failure rates corresponding to these limits are

1/1 844 and 1/637, respectively, i.e., 0·000 542 4 and 0·001 57. Thus we may assume the failure rate to lie between 0·000 542 4 and 0·001 57 at 90 per cent level of confidence.

9. In a test to determine failure rate/stress curves for derating purposes, three sets of observations were made:

   (a) failure rate for rated voltage and temperature,
   (b) failure rate for rated voltage and twice rated temperature,
   (c) failure rate for half rated voltage and rated temperature.

If the failure rates for these conditions are 0·001, 0·004, and 0·000 05 respectively, and the rated temperature is 20°C, calculate the probable failure rate at one-third rated voltage and one-third rated temperature, assuming proportionality ratios are constant as in eq. 41.

From eq. 41, for test 2

$$0·004 = 0·001 \times K^{20};$$

thus

$$K^{20} = 4$$

and

$$\ln K = \tfrac{1}{20} \ln 4.$$

Hence

$$K = 1·072.$$

For test 3

$$0·000\ 05 = 0·001(\tfrac{1}{2})^n \quad \text{and} \quad 2^n = 20;$$

hence

$$n = \ln 20/\ln 2 = 4·32.$$

The equation may thus be assumed to be

$$\lambda_x = 0·001 \left(\frac{V_x}{V_m}\right)^{4·32} 1·072^{t_x - t_m}$$

57

and for $\frac{1}{3}$ rated voltage and $\frac{1}{3}$ rated temperature

$$\lambda_x = \frac{0\cdot001 \times 1\cdot072^{-40/3}}{3^{4\cdot32}}$$

$$= \frac{0\cdot001}{\sqrt[3]{1\cdot072^{40}} \times 3^{4\cdot32}} = 0\cdot000\ 003\ 439.$$

The dramatic improvement in failure rate obtainable by derating is clearly shown.

Summary of relevant formulae

For a constant proportional failure rate,

$$R = e^{-\lambda t} \tag{1}$$

$$R + Q = 1 \tag{2}$$

$$Q = 1 - e^{-\lambda t} \tag{3}$$

$$N_s = N_0\, e^{-\lambda t} \tag{4}$$

$$N_f = N_0\,(1 - e^{-\lambda t}) \tag{5}$$

$$m = 1/\lambda \tag{6}$$

$$R = e^{-t/m} \tag{7}$$

$$Q = 1 - e^{-t/m} \tag{8}$$

$$N_s = N_0\, e^{-t/m} \tag{9}$$

$$N_f = N_0\,(1 - e^{-t/m}) \tag{10}$$

where R denotes reliability
Q denotes unreliability
λ denotes proportional failure rate (i.e., failure rate expressed as a proportion of N_0)
N_s denotes number of live components (survivors)
N_f denotes number of dead components (failures)
N_0 denotes initial number of live components
m denotes mean time between (chance) failures
t denotes time.

For a test to determine MTBF, m

$$m = \frac{\text{total survival hours}}{\text{number of failures}}. \qquad (11)$$

For a normally distributed variable x, the standard deviation σ is given by

$$\sigma = \sqrt{\frac{\Sigma \, (x - x_m)^2}{n}} \qquad (12)$$

where x_m is the mean value of n observations of x.

In a test to determine mean wearout life M,

$$M = \frac{\text{sum of lives}}{\text{number of components}} \qquad (13)$$

For an exponentially distributed variable,

$$\text{upper confidence limit} = \frac{2nm}{\chi^2_{1-\alpha/2,\,n}} \qquad (14)$$

$$\text{lower confidence limit} = \frac{2nm}{\chi^2_{\alpha/2,\,n}} \qquad (15)$$

at a level of confidence given by $100(1 - \alpha)$ per cent,

where n denotes number of failures

        m denotes an estimate of the mean value of the variable

        χ^2 denotes the value of chi-squares (given in tables) for values of n and α or $(1 - \alpha/2)$.

For a component or unit which forms part of a system

$$\frac{m_c}{m_s} = \frac{t_c}{t_s} \qquad (16)$$

$$m_s = \frac{m_c}{d} \qquad (17)$$

where m_c denotes component MTBF in component operating hours

m_s denotes component MTBF in system operating hours

t_c denotes component operating hours

t_s denotes system operating hours

d denotes duty cycle $(d = t_c/t_s)$.

The probability of both events x and y occurring, P_{xy}, is given by

$$P_{xy} = P_x P_y \qquad (18)$$

and the probability of either event x or event y occurring, P_{x+y}, is given by

$$P_{x+y} + P_x + P_y - P_x P_y \qquad (19)$$

where P_x denotes the probability of x occurring

P_y denotes the probability of y occurring.

The following equations refer to series and parallel systems. The symbols used are as above for reliability, unreliability, failure rate, etc., with the addition of appropriate subscripts as follows:

subscript s denotes series

subscript p denotes parallel

subscripts 1, 2, 3, etc., denote components or subunits 1, 2, 3, etc.

$$R_s = R_1 R_2 \cdots \qquad (20)$$

$$Q_s = Q_1 + Q_2 - Q_1 Q_2 \qquad (21)$$

$$Q_s = 1 - R_s \qquad (22)$$

$$R_p = R_1 + R_2 - R_1 R_2 \qquad (23)$$

$$Q_p = Q_1 Q_2 \qquad (24)$$

$$Q_p = 1 - R_p \qquad (25)$$

$$R_s = e^{-(\lambda_1 + \lambda_2 + \lambda_3 + \cdots)t} \qquad (26)$$

$$\lambda_s = \lambda_1 + \lambda_2 + \lambda_3 + \cdots \qquad (27)$$

$$m_s = 1/\lambda_s \qquad (28)$$

$$R_s = e^{-n\lambda t} \qquad (29)$$

$$\lambda_s = n\lambda \qquad (30)$$

$$m_s = 1/(n\lambda) \qquad (31)$$

$$R_p = R_1 + R_2 - R_1 R_2 \qquad (32)$$

$$R_p = R_1 + R_2 + R_3 - R_1 R_2 - R_2 R_3 \\ - R_1 R_3 + R_1 R_2 R_3 \qquad (33)$$

$$R_p = e^{-\lambda_1 t} + e^{-\lambda_2 t} + e^{-(\lambda_1 + \lambda_2)t} \qquad (34)$$

$$R_p = e^{-\lambda_1 t} + e^{-\lambda_2 t} + e^{-\lambda_3 t} + 3^{-(\lambda_1 + \lambda_2)t} \\ e^{-(\lambda_2 + \lambda_3)t} + e^{-(\lambda_1 + \lambda_3)t} + e^{-(\lambda_1 + \lambda_2 + \lambda_3)t} \qquad (35)$$

$$m_p = \frac{1}{\lambda_1} + \frac{1}{\lambda_2} - \frac{1}{\lambda_1 + \lambda_2} \qquad (36)$$

$$m_p = \frac{1}{\lambda_1} + \frac{1}{\lambda_2} + \frac{1}{\lambda_3} - \frac{1}{\lambda_1 + \lambda_2} - \frac{1}{\lambda_2 + \lambda_3} \\ - \frac{1}{\lambda_1 + \lambda_3} + \frac{1}{\lambda_1 + \lambda_2 + \lambda^3} \qquad (37)$$

$$m_p = \frac{1}{\lambda} + \frac{1}{2\lambda} + \frac{1}{3\lambda} + \cdots + \frac{1}{n\lambda} \qquad (38)$$

where n in eqs. (30), (31), and (38) denotes the number of components or subunits having equal failure rates.

For a system, the utilization factor U is given by

$$U = \frac{\text{operating time}}{\text{maintenance time} + \text{idle time} + \text{operating time}} \qquad (39)$$

the availability (maximum utilization factor) A is given by

$$A = U_{\max}$$

$$= \frac{\text{operating time}}{\text{minimum maintenance time} + \text{operating time}}. \qquad (40)$$

For any two sets of operating conditions denoted by x and m respectively, the voltages, V_x and V_m, temperatures, t_x and t_m, and failure rates, λ_x and λ_m are related by the equation

$$\lambda_x = \lambda_m \left(\frac{V_x}{V_m}\right)^n K^{t_x - t_m}, \tag{41}$$

where n and K are constants over a limited range of conditions and may be determined by the equations

$$K = \left(\frac{\lambda_x}{\lambda_m}\right) \frac{1}{t_x - t_m} \tag{42}$$

for a constant voltage test, and

$$n = \frac{\ln (\lambda_x / \lambda_m)}{\ln (V_m / V_x)} \tag{43}$$

for a constant temperature test.

Self-test questions and problems

The answers to descriptive questions are contained in the text; the answers to numerical questions are alongside in parentheses.

1. Define the terms *reliability* and *failure* as applied to a component or system subunit. Explain the following categories of failure:

(a) misuse   (d) gradual
(b) inherent weakness   (e) partial
(c) sudden   (f) complete

(g) catastrophic   (j) chance
(h) degradation   (k) wearout

2. Define (a) instantaneous failure rate, (b) proportional failure rate, and (c) percentage failure rate.

3. Sketch a typical bathtub diagram and indicate the burn-in period, constant failure rate period and wearout failure period.

4. Calculate the probable number of survivors out of a batch of 1 000 components after 1 000 hours of operation if the failure rate may be assumed constant at 4 per cent/ 1 000 hours.   (961)

5. Determine the percentage unreliability of the components of question (4) over a period of 2 000 hours.
                                              (7·7 per cent)

6. Define MTBF. It is desired to achieve a reliability of 0·95 for a certain component over a period of 500 hours. calculate the necessary MTBF.   (9 672 hours)

7. In a test lasting 100 hours, to determine the MTBF for a certain component, 7 out of 50 failed, one after each of the following periods (in hours): 5, 15, 16, 25, 32, 34, and 81. Estimate the MTBF and hence the anticipated number of failures out of a batch of 1 000 if the test is run over 2 000 hours. Wearout may be ignored.
                                        (644 hours, 958)

8. The earth-vehicle communications system of a manned lunar craft has an MTBF of 10 000 hours. Estimate the probability that contact will be maintained during a mission lasting 200 hours.   (98 per cent)

9. Define the term *mean wearout life* as applied to reliability studies.

64

A wearout test gave the following data:

Number of components	1	1	2	4	1	1
Life (hours)	3 000	3 100	3 050	2 900	2 950	2 990

Calculate the mean wearout life of these components.

(2 974 hours)

10. Describe what is meant by *confidence level* and *confidence limits* and how these may be determined for (a) a normally distributed variable, and (b) an exponentially distributed variable.

11. The data for a certain wearout test is given below:

Number of components	1	4	7	8	5	3
Life per component (hours)	1 400	800	900	1 000	1 100	1 200

Calculate the upper and lower confidence limits for the mean wearout life at a 75 per cent level of confidence.

(969·3 hours, 1 030·7 hours)

12. Calculate the upper and lower confidence limits of the MTBF of a certain component at a 95 per cent confidence level when a test during which there were 20 failures gave an estimate of 100 hours.

(67 hours, 164 hours)

13. Briefly explain the effect on the reliability of a component when it is included in a system which requires only intermittent operation of the component for full operation of the system.

The failure rate of a certain component may be assumed constant at 0·000 1 per hour. If the component forms part of a system, in which it operates for only 25 per cent of the system operating time, calculate the component MTBF in system hours. (40 000 hours)

14. Define the laws of probability for two independent events occurring (a) either singly or together, and (b) together.

If two dice are thrown together calculate the probability of (a) a 5 turning up on either die, (b) a 4 and a 5 turning up in a single throw. $\left(\frac{11}{36};\frac{1}{36}\right)$

15. Describe the difference between a series and parallel system in the context of reliability considerations and state which may be expected to have the higher reliability for the same number of subunits.

16. A broadcast relay system relies on ten repeaters each of which must be in working order for reception to be obtained at the distributing station. The overall reliability for 1 000 hours operation of each repeater is calculated as 0·89. Determine the MTBF for the system neglecting wearout failures.   (858·2 hours)

17. Define the terms *maintainability, availability,* and *dependability.* Calculate the ratio between maintenance and operating time necessary to achieve an availability of 0·75 for any system.   (1:3)

18. Assuming a ratio between maintenance and operating time of 1:5 calculate the ratio between the maximum time a system can be allowed to stand idle and the operating time to achieve a minimum utilization factor of 0·8.   (0·05:1)

19. Discuss the possible stresses acting on any component categorized into (a) operating and (b) environmental stresses. Include in the discussion mention of methods of stress reduction and the meaning of the term *derating.*

20. Assuming that the fifth power law for voltage-induced stress is valid and that failure rate doubles for every 10°C rise in temperature, calculate the percentage reduction in failure rate for operation at 50 per cent rated voltage and 50 per cent rated temperature, assuming a rated temperature of 20°C. (93·7 per cent)

Bibliography

I. Bazovsky, *Reliability Theory and Practice*, Prentice-Hall (1962).

Dummer and Griffin, *Electronics Reliability–Calculation and Design*, Pergamon Press (1965).

Loveday, *Statistics, a Second Course*, Cambridge University Press (1961).

Appendix

Some typical reliability figures for electronic components

Failure rate (per cent/1 000 hours)†

Transistors
silicon	0·04
germanium	0·1
Triode valves	0·1

† It must be emphasized that test figures vary greatly. The ones given are for operation in approximately similar conditions.

Resistors
 composition   0·004
 film   0·19
 wirewound   0·03
Capacitors
 paper   0·05
 mica   0·004
 ceramic   0·025
 electrolytic   0·98
Integrated circuits   less than 0·01

Index

Made and printed by offset in Great Britain
by William Clowes and Sons, Limited
London and Beccles